GUESS
COUNTRY
SILHOUETTE

How many of the 196 country silhouettes can you recognise?

B.C. Lester Books

A QUICK MESSAGE FROM THE PUBLISHER...

THANKS FOR PURCHASING THIS BOOK...

...we really hope you enjoy it. If you have the chance, then all feedback on Amazon is greatly appreciated. We have put a lot of effort into making this book, so if you are not completely satisfied, please email us at ben@bclesterbooks.com and we will do our best to address any issues. If you have any suggestions, want to get in touch or want to send us your score, then email at the same address - ben@bclesterbooks.com

IS THIS BOOK MISPRINTED?
Printing presses, like humans, aren't quite perfect. Send us an email at ben@bclesterbooks.com with a photo of the misprint, and we will get another copy sent out to you!

WHO ARE WE AT B.C. LESTER BOOKS?

B.C. Lester Books is a small publishing firm of three people based in Buckinghamshire, UK. We aim to provide quality works in all things geography, for kids and adults, with varying interests. We have already released a selection of activity, trivia and fact books and are working hard to bring you wider selection. Have a suggestion for us? Then email ben@bclesterbooks.com. We are all ears!

HAVE EVEN MORE QUIZZING FUN WITH OUR GIFT TO YOU: A 3-IN-1 GEOGRAPHY QUIZ BOOK!

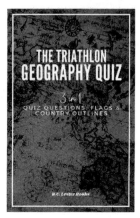

Go here to grab your FREE copy!
www.bclesterbooks.com/freebies/

THE COUNTRY SILHOUETTE QUIZ!

GOOD LUCK!

NOTES:

On each double page you will find clues for the countries of the previous 4 countries.

Answers begin at Page 104.

The scorecard is at Page 107.

PAGE 5
For clues, see reverse page. Answers on Page 104

SCORE

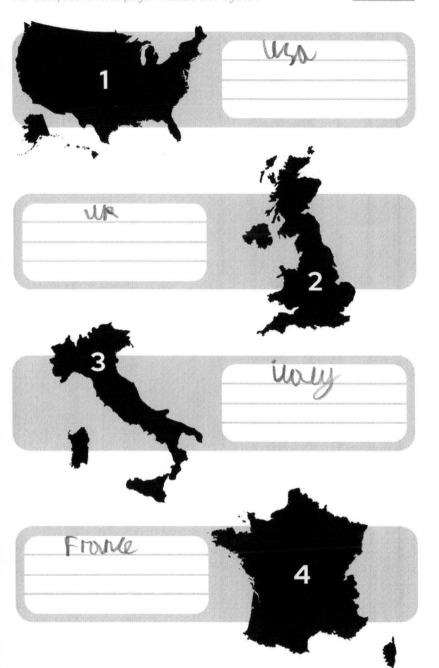

1. USA

2. UK

3. Italy

4. France

CLUES:

1 U

2 U

3 I

4 F

PAGE 7

For clues, see reverse page. Answers on Page 104

SCORE

5

6

7

8

CLUES:

5 C

6 A

7 M

8 C

PAGE 9
For clues, see reverse page. Answers on Page 104

SCORE

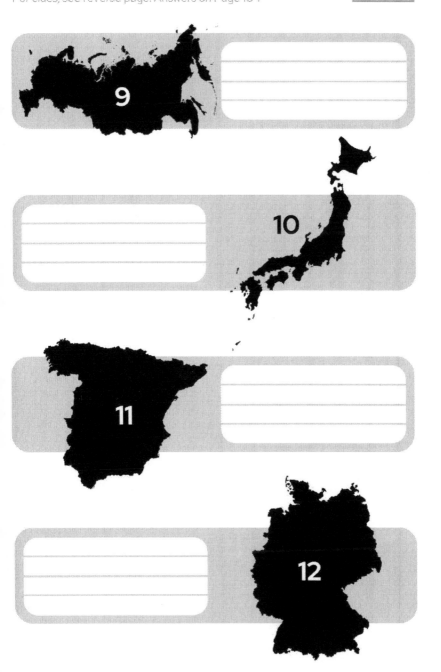

CLUES:

9 R

10 J

11 S

12 G

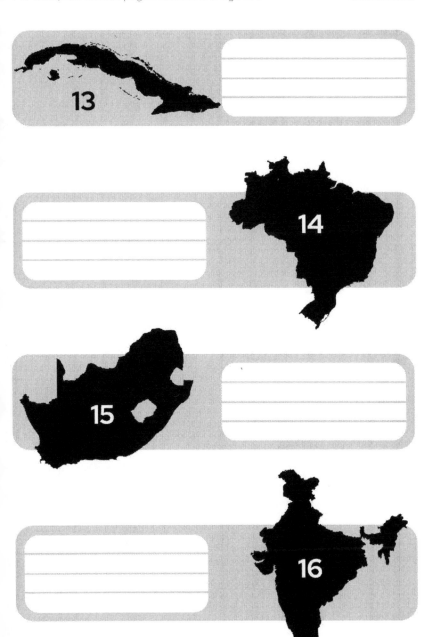

CLUES:

13 C

14 B

15 S

16 I

PAGE 13
For clues, see reverse page. Answers on Page 104

SCORE

17

18

19

20

CLUES:

17 G

18 C

19 I

20 N

For clues, see reverse page. Answers on Page 104

SCORE

21

22

23

24

CLUES:

21 I

22 A

23 N

24 S

For clues, see reverse page. Answers on Page 104

SCORE

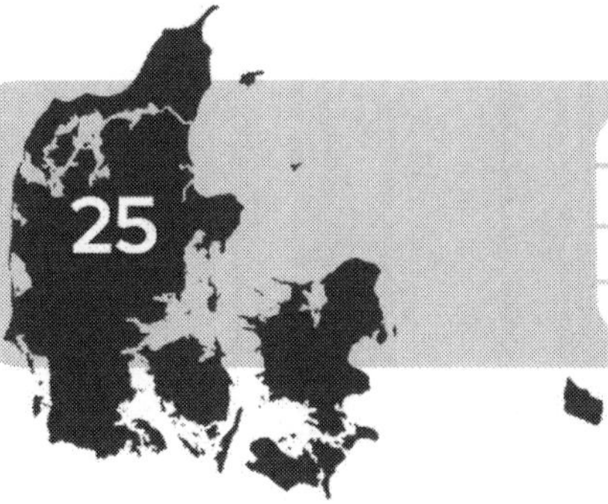

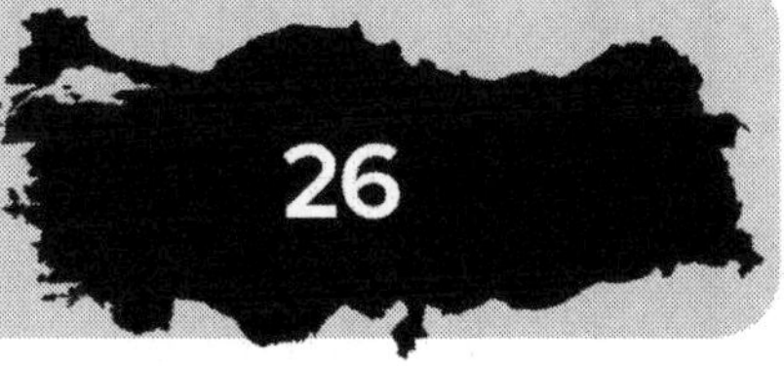

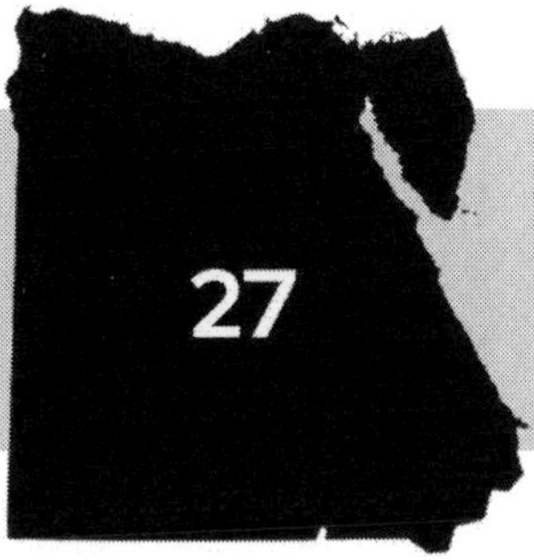

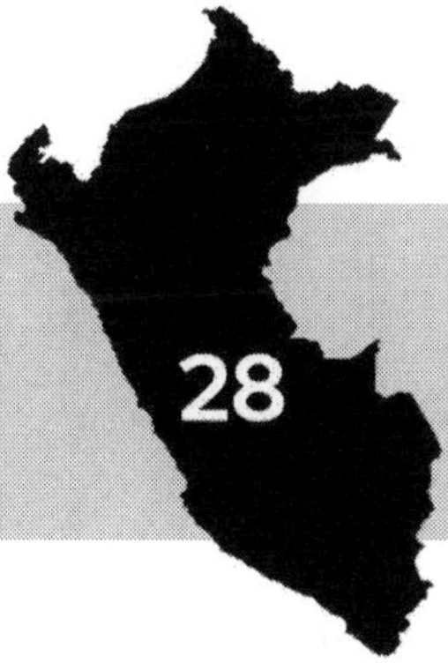

CLUES:

25 D

26 T

27 E

28 P

PAGE 19
For clues, see reverse page. Answers on Page 104

SCORE

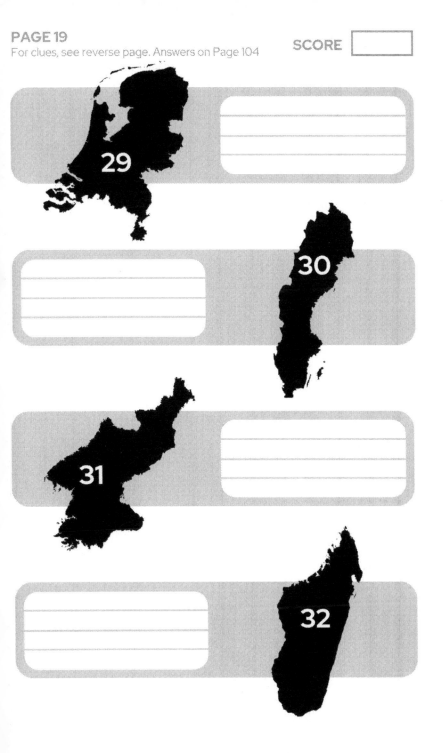

CLUES:

29 N

30 S

31 N

32 M

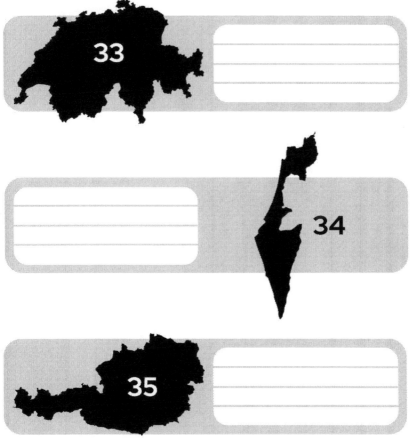

CLUES:

33 S

34 I

35 A

36 N

PAGE 23
For clues, see reverse page. Answers on Page 104

SCORE

37

38

39

40

CLUES:

37 S

38 I

39 T

40 C

PAGE 25
For clues, see reverse page. Answers on Page 104

SCORE

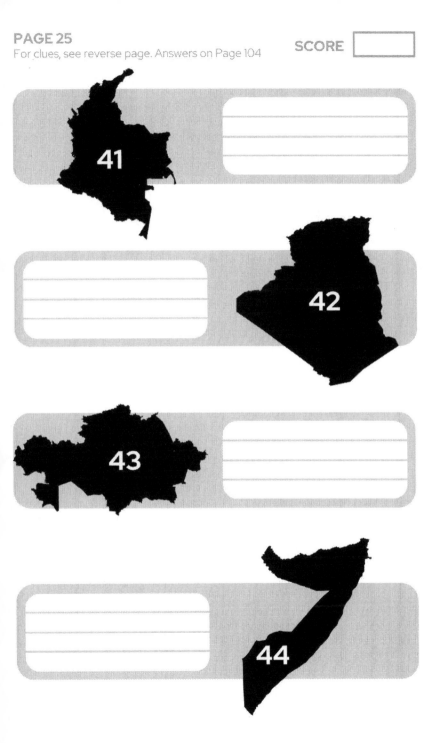

CLUES:
41 C
42 A
43 K
44 S

PAGE 27
For clues, see reverse page. Answers on Page 104

SCORE

45

46

47

48

CLUES:

45 R

46 K

47 I

48 B

PAGE 29
For clues, see reverse page. Answers on Page 104

SCORE

49

50

51

52

CLUES:

49 B

50 S

51 U

52 V

PAGE 31
For clues, see reverse page. Answers on Page 104

SCORE

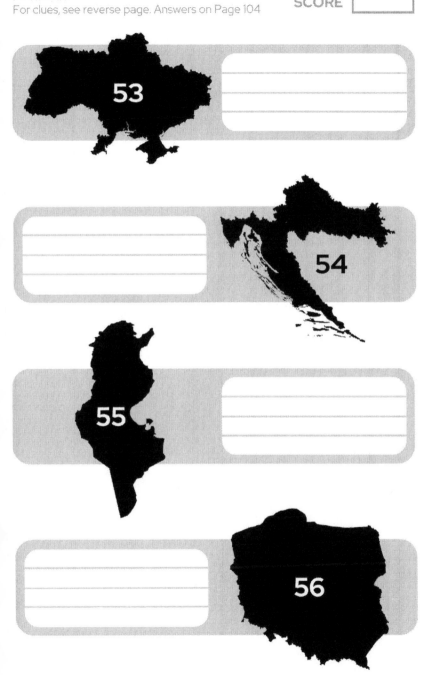

CLUES:

53 U

54 C

55 T

56 P

PAGE 33
For clues, see reverse page. Answers on Page 104

SCORE

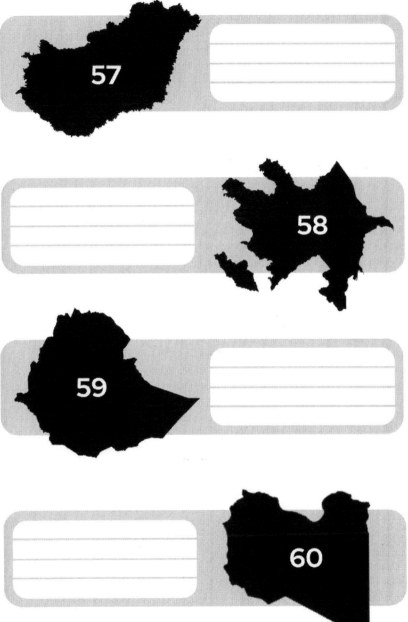

CLUES:

57 H

58 A

59 E

60 L

For clues, see reverse page. Answers on Page 104

SCORE

61

62

63

64

CLUES:
61 F
62 G
63 B
64 M

SCORE

65

66

67

68

CLUES:

65 A

66 M

67 P

68 V

PAGE 39
For clues, see reverse page. Answers on Page 105

SCORE

69

70

71

72

CLUES:
69 J
70 S
71 D
72 T

SCORE

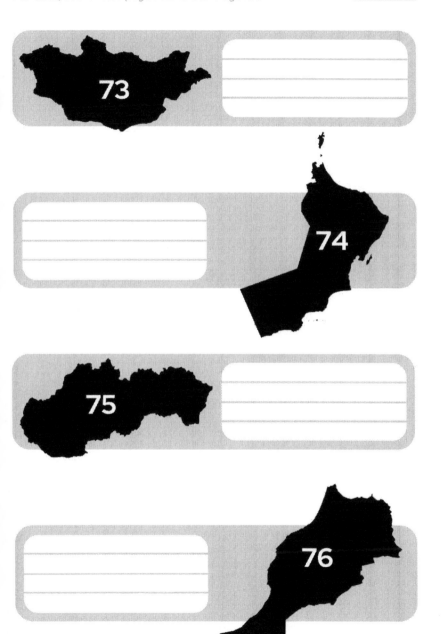

CLUES:

73 M

74 O

75 S

76 M

PAGE 43
For clues, see reverse page. Answers on Page 105

SCORE

77

78

79

80

CLUES:

77 P

78 C

79 P

80 I

PAGE 45
For clues, see reverse page. Answers on Page 105

SCORE

81

82

83

84

CLUES:

81 P

82 A

83 H

84 P

PAGE 47
For clues, see reverse page. Answers on Page 105

SCORE

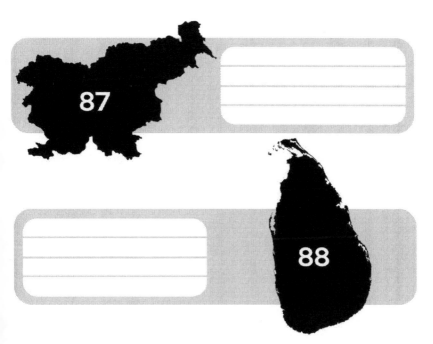

CLUES:
85 D
86 C
87 S
88 S

PAGE 49
For clues, see reverse page. Answers on Page 105

SCORE

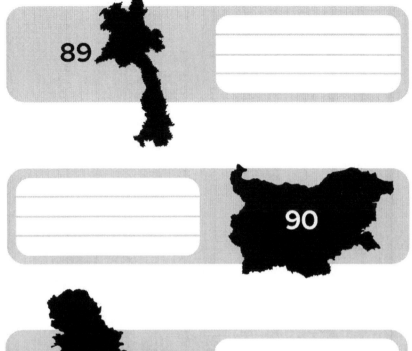

CLUES:
89 L
90 B
91 S
92 B

PAGE 51
For clues, see reverse page. Answers on Page 105

SCORE

93

94

95

96

CLUES:
93 T
94 C
95 L
96 P

PAGE 53
For clues, see reverse page. Answers on Page 105

SCORE

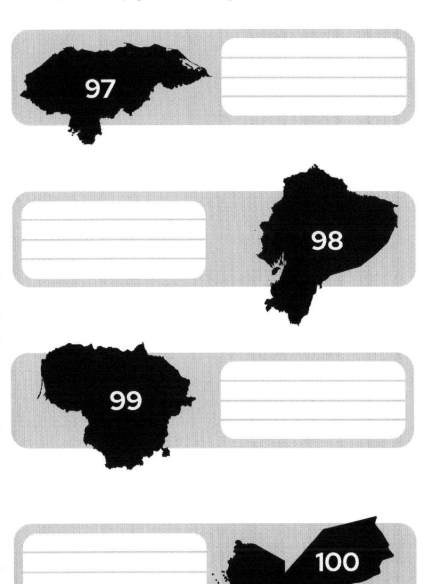

CLUES:

97 H

98 E

99 L

100 Y

PAGE 55
For clues, see reverse page. Answers on Page 105

SCORE

101

102

103

104

CLUES:
101 N
102 M
103 E
104 N

SCORE

105

106

107

108

CLUES:
105 L
106 P
107 M
108 G

SCORE

109

110

111

112

CLUES:
109 S
110 A
111 C
112 T

SCORE

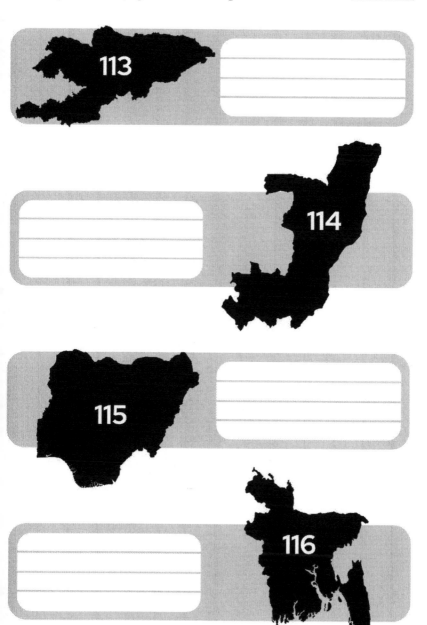

113

114

115

116

CLUES:
113 K
114 R
115 N
116 B

SCORE

117

118

119

120

CLUES:
117 G
118 E
119 A
120 M

PAGE 65
For clues, see reverse page. Answers on Page 105

SCORE

121

122

123

124

CLUES:

121 M
122 N
123 A
124 U

SCORE

125

126

127

128

CLUES:
125 B
126 M
127 S
128 T

For clues, see reverse page. Answers on Page 106

SCORE

129

130

131

132

CLUES:
129 M
130 S
131 G
132 C

SCORE

133

134

135

136

CLUES:

133 U

134 S

135 G

136 K

For clues, see reverse page. Answers on Page 106

SCORE

137

138

139

140

CLUES:
137 B
138 L
139 E
140 B

For clues, see reverse page. Answers on Page 106

SCORE

141

142

143

144

CLUES:
141 Z
142 B
143 U
144 C

PAGE 77
For clues, see reverse page. Answers on Page 106

SCORE

145

146

147

148

CLUES:
145 T
146 B
147 S
148 I

SCORE

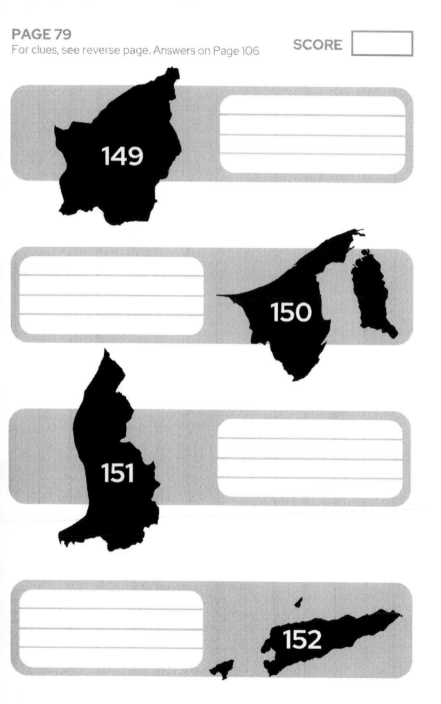

CLUES:
149 S
150 B
151 L
152 E

SCORE

153

154

155

156

CLUES:

153 G

154 E

155 G

156 G

PAGE 83
For clues, see reverse page. Answers on Page 106

SCORE

CLUES:
157 C
158 M
159 B
160 P

PAGE 85
For clues, see reverse page. Answers on Page 106

SCORE

161

162

163

164

CLUES:
161 Q
162 K
163 S
164 S

SCORE

165

166

167

168

CLUES:

165 B

166 C

167 L

168 B

PAGE 89

For clues, see reverse page. Answers on Page 106

SCORE

169

170

171

172

CLUES:
169 E
170 L
171 D
172 D

PAGE 91
For clues, see reverse page. Answers on Page 106

SCORE

173

174

175

176

CLUES:

173 A

174 V

175 K

176 G

PAGE 93
For clues, see reverse page. Answers on Page 106

SCORE

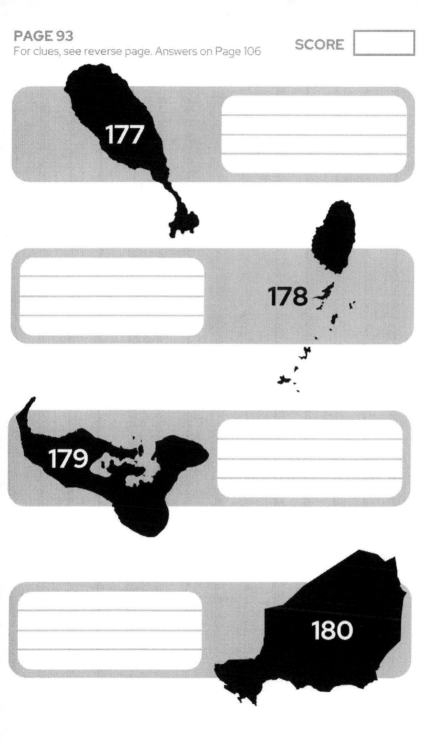

177

178

179

180

CLUES:
177 S
178 S
179 T
180 N

PAGE 95
For clues, see reverse page. Answers on Page 106

SCORE

181

182

183

184

CLUES:
181 B
182 F
183 M
184 M

PAGE 97
For clues, see reverse page. Answers on Page 106

SCORE

185

186

187

188

CLUES:
185 N
186 T
187 S
188 S

SCORE

189

190

191

192

CLUES:
189 Z
190 M
191 M
192 S

SCORE

CLUES:

193 J

194 M

195 R

196 V

THE
ANSWERS

ANSWERS

1 UNITED STATES
2 UNITED KINGDOM
3 ITALY
4 FRANCE
5 CANADA
6 AUSTRALIA
7 MEXICO
8 CHINA
9 RUSSIA
10 JAPAN
11 SPAIN
12 GERMANY
13 CUBA
14 BRAZIL
15 SOUTH AFRICA
16 INDIA
17 GREECE
18 CHILE
19 INDONESIA
20 NEW ZEALAND
21 IRAN
22 ARGENTINA
23 NORWAY
24 SOUTH KOREA
25 DENMARK
26 TURKEY
27 EGYPT
28 PERU
29 NETHERLANDS
30 SWEDEN
31 NORTH KOREA
32 MADAGASCAR
33 SWITZERLAND
34 ISRAEL
35 AUSTRIA
36 NEPAL
37 SAUDI ARABIA
38 ICELAND
39 THAILAND
40 CYPRUS
41 COLOMBIA
42 ALGERIA
43 KAZAKHSTAN
44 SOMALIA
45 ROMANIA
46 KENYA
47 IRAQ
48 BOLIVIA
49 BELGIUM
50 SYRIA
51 UNITED ARAB
EMIRATES
52 VIETNAM
53 UKRAINE
54 CROATIA
55 TUNISIA
56 POLAND
57 HUNGARY
58 AZERBAIJAN
59 ETHIOPIA
60 LIBYA
61 FINLAND
62 GUATEMALA
63 THE BAHAMAS
64 MYANMAR

65 AFGHANISTAN	97 HONDURAS
66 MALAYSIA	98 ECUADOR
67 PANAMA	99 LITHUANIA
68 VENEZUELA	100 YEMEN
69 JORDAN	101 NICARAGUA
70 SUDAN	102 MOZAMBIQUE
71 DR CONGO	103 ESTONIA
72 TRINIDAD AND	104 NORTH MACEDONIA
TOBAGO	105 LUXEMBOURG
73 MONGOLIA	106 PALESTINE
74 OMAN	107 MALTA
75 SLOVAKIA	108 GUYANA
76 MOROCCO	109 SURINAME
77 PHILIPPINES	110 ANDORRA
78 CAMBODIA	111 CHAD
79 PAKISTAN	112 TURKMENISTAN
80 IRELAND	113 KYRGYZSTAN
81 PORTUGAL	114 REPUBLIC OF THE
82 ALBANIA	CONGO
83 HAITI	115 NIGERIA
84 PAPUA NEW GUINEA	116 BANGLADESH
85 DOMINICAN REPUBLIC	117 GHANA
86 CZECH REPUBLIC	118 ERITREA
87 SLOVENIA	119 ARMENIA
88 SRI LANKA	120 MONTENEGRO
89 LAOS	121 MOLDOVA
90 BULGARIA	122 NAMIBIA
91 SERBIA	123 ANGOLA
92 BOSNIA &	124 URUGUAY
HERZEGOVINA	125 BELARUS
93 TANZANIA	126 MAURITANIA
94 COSTA RICA	127 SINGAPORE
95 LATVIA	128 TOGO
96 PARAGUAY	

129 MONACO
130 SENEGAL
131 THE GAMBIA
132 CAMEROON
133 UGANDA
134 SOUTH SUDAN
135 GEORGIA
136 KOSOVO
137 BELIZE
138 LIBERIA
139 EL SALVADOR
140 BOTSWANA
141 ZAMBIA
142 BHUTAN
143 UZBEKISTAN
144 CENTRAL AFRICAN REPUBLIC
145 TAJIKISTAN
146 BURKINA FASO
147 SIERRA LEONE
148 IVORY COAST / COTE D'IVOIRE
149 SAN MARINO
150 BRUNEI
151 LIECHTENSTEIN
152 EAST TIMOR / TIMOR LESTE
153 GABON
154 EQUATORIAL GUINEA
155 GUINEA
156 GUINEA-BISSAU
157 CAPE VERDE
158 MAURITIUS
159 BARBADOS
160 PALAU

161 QATAR
162 KUWAIT
163 SAINT LUCIA
164 SEYCHELLES
165 BAHRAIN
166 COMOROS
167 LEBANON
168 BURUNDI
169 ESWATINI
170 LESOTHO
171 DJIBOUTI
172 DOMINICA
173 ANTIGUA AND BARBUDA
174 VANUATU
175 KIRIBATI
176 GRENADA
177 SAINT KITTS AND NEVIS
178 SAINT VINCENT AND THE GRENADINES
179 TONGA
180 NIGER
181 BENIN
182 FIJI
183 MALDIVES
184 MALI
185 NAURU
186 TUVALU
187 SAMOA
188 SOLOMON ISLANDS
189 ZIMBABWE
190 MALAWI
191 MICRONESIA
192 SAO TOME AND PRINCIPE
193 JAMAICA
194 MARSHALL ISLANDS
195 RWANDA
196 VATICAN CITY

SCORECARD TOTAL [/196]

1-4	/4	65-68	/4	129-132	/4
5-8	/4	69-72	/4	133-136	/4
9-12	/4	73-76	/4	137-140	/4
13-16	/4	77-80	/4	141-144	/4
17-20	/4	81-84	/4	145-148	/4
21-24	/4	85-88	/4	149-152	/4
25-28	/4	89-92	/4	153-156	/4
29-32	/4	93-96	/4	157-160	/4
33-36	/4	97-100	/4	161-164	/4
37-40	/4	101-104	/4	165-168	/4
41-44	/4	105-108	/4	169-172	/4
45-48	/4	109-112	/4	173-176	/4
49-52	/4	113-116	/4	177-180	/4
53-56	/4	117-120	/4	181-184	/4
57-60	/4	121-124	/4	185-188	/4
61-64	/4	125-128	/4	189-196	/8

SCORECARDS
OF 10 GEOGRAPHY WHIZZES

1 190 – B.C.

2 186 – C.S.

3 180 – J.L. + H.H.

5 176 – J.O.

6 174 – A.P.

7 164 – J.G.

8 150 – K.T.

9 148 – I.Z.

10 132 – A.M.

Printed in Great Britain
by Amazon